WE ARE ALL TIME
TRAVELLERS,
JOURNEYING
TOGETHER INTO
THE FUTURE. BUT
LET US WORK
TOGETHER TO
MAKE THAT FUTURE
A PLACE WE
WANT TO VISIT

What Is Inside a Black Hole?

STEPHEN HAWKING was a brilliant theoretical physicist and is generally considered to have been one of the world's greatest thinkers. He held the position of Lucasian Professor of Mathematics at the University of Cambridge for thirty years and is the author of *A Brief History of Time*, which was an international bestseller. His other books for the general reader include *A Briefer History of Time*, the essay collection *Black Holes and Baby Universes*, *The Universe in a Nutshell*, *The Grand Design* and *Black Holes: The BBC Reith Lectures*. He died on 14 March 2018.

'What is inside a black hole?' and 'Is time travel possible?' are essays taken from Stephen Hawking's final book, *Brief Answers to the Big Questions* (John Murray, 2018).

STEPHEN HAWKING

What Is Inside a Black Hole?

JOHN MURRAY

First published in Great Britain in 2022 by John Murray (Publishers)
An Hachette UK company

8

Copyright © Spacetime Publications Limited 2018

The right of Stephen Hawking to be identified as the
Author of the Work has been asserted by him in accordance
with the Copyright, Designs and Patents Act 1988.

'What is inside a black hole?' and 'Is time travel possible?'
are essays taken from *Brief Answers to the Big Questions*,
published by John Murray (2018)

A CIP catalogue record for this title
is available from the British Library

Paperback ISBN 978-1-529-39236-4
eBook ISBN 978-1-529-39237-1

Text design by Craig Burgess

Typeset in Sabon MT by
Palimpsest Book Production Ltd, Falkirk, Stirlingshire

Printed and bound in Great Britain by Clays Ltd, Elcograf S.p.A.

John Murray policy is to use papers that are natural, renewable and
recyclable products and made from wood grown in sustainable forests.
The logging and manufacturing processes are expected to conform
to the environmental regulations of the country of origin.

John Murray (Publishers)
Carmelite House
50 Victoria Embankment
London EC4Y 0DZ

www.johnmurraypress.co.uk

Contents

WHAT IS INSIDE A BLACK HOLE?....1

IS TIME TRAVEL POSSIBLE? 37

WHAT IS INSIDE
A BLACK HOLE?

IT is said that fact is sometimes stranger than fiction, and nowhere is that more true than in the case of black holes. Black holes are stranger than anything dreamed up by science-fiction writers, but they are firmly matters of science fact.

The first discussion of black holes was in 1783, by a Cambridge man, John Michell. His argument ran as follows. If one fires a particle, such as a cannon ball, vertically upwards, it will be slowed down by gravity. Eventually, the particle will stop moving upwards, and will fall back. However, if the initial upwards velocity were greater than some critical value, called the escape velocity, gravity would never be strong enough to stop the particle, and it would get away. The escape velocity is just over 11 kilometres per second for the Earth, and about 617 kilometres per second for the Sun. Both of these are much higher than the speed of real cannon balls. But they are low compared to the speed of light, which is 300,000 kilometres per second. Thus light can get away from the Earth or Sun without much difficulty. However, Michell argued that there could be stars that were much more massive than the Sun which had escape velocities greater than the speed of light. We would not be able to see them, because any light they sent out would be dragged back by gravity. Thus they would be what Michell called dark stars, what we now call black holes.

To understand them, we need to start with gravity. Gravity is described by Einstein's general theory of relativity, which is a theory of space and time as well as gravity. The behaviour of space and time is governed by a set of equations called the Einstein equations which Einstein put forward in 1915. Although gravity is by far the weakest of the known forces of nature, it has two crucial advantages over other forces. First, it acts over a long range. The earth is held in orbit by the Sun, ninety-three million miles away, and the Sun is held in orbit around the centre of the galaxy, about 10,000 light years away. The second advantage is that gravity is always attractive, unlike electric forces which can be either attractive or repulsive. These two features mean that for a sufficiently large star the gravitational attraction between particles can dominate over all other forces and lead to gravitational collapse. Despite these facts, the scientific community was slow to realise that massive stars could collapse in on themselves under their own gravity and to figure out how the object left behind would behave. Albert Einstein even wrote a paper in 1939 claiming that stars could not collapse under gravity, because matter could not be compressed beyond a certain point. Many scientists shared Einstein's gut feeling. The principal exception was the American scientist John Wheeler, who in many ways is the hero of the black hole story. In his work in the 1950s and 1960s, he emphasised that many stars would eventually collapse, and explored the problems this posed for theoretical physics. He also foresaw many of the properties of the objects which collapsed stars become – that is, black holes.

During most of the life of a normal star, over many billions of years, it will support itself against its own gravity by thermal pressure caused by nuclear processes which convert hydrogen into helium. Eventually, however, the star will exhaust its nuclear fuel.

The star will contract. In some cases, it may be able to support itself as a white dwarf star, the dense remnants of a stellar core. However, Subrahmanyan Chandrasekhar showed in 1930 that the maximum mass of a white dwarf star is about 1.4 times that of the Sun. A similar maximum mass was calculated by the Russian physicist Lev Landau for a star made entirely of neutrons.

What would be the fate of those countless stars with a greater mass than the maximum mass of a white dwarf or neutron star once they had exhausted nuclear fuel? The problem was investigated by Robert Oppenheimer of later atom bomb fame. In a couple of papers in 1939, with George Volkoff and Hartland Snyder, he showed that such a star could not be supported by pressure. And that if one neglected pressure a uniform spherically systematic symmetric star would contract to a single point of infinite density. Such a point is called a singularity. All our theories of space are formulated on the assumption that space-time is smooth and nearly flat, so they break down at the singularity, where the curvature of space-time is infinite. In fact, it marks the end of space and time itself. That is what Einstein found so objectionable.

Then the Second World War intervened. Most scientists, including Robert Oppenheimer, switched their attention to nuclear physics, and the issue of gravitational collapse was largely forgotten. Interest in the subject revived with the discovery of distant objects called quasars. The first quasar, 3C273, was found in 1963. Many other quasars were soon discovered. They were bright despite being at great distances from the Earth. Nuclear processes could not account for their energy output, because they release only a small fraction of their rest mass as pure energy. The only alternative was gravitational energy released by gravitational collapse.

Gravitational collapse of stars was rediscovered. When this happens, the gravity of the object draws all its surrounding matter inwards. It was clear that a uniform spherical star would contract to a point of infinite density, a singularity. But what would happen if the star isn't uniform and spherical? Could this unequal distribution of the star's matter cause a non-uniform collapse and avoid a singularity? In a remarkable paper in 1965, Roger Penrose showed there would still be a singularity, using only the fact that gravity is attractive.

The Einstein equations can't be defined at a singularity. This means that at this point of infinite density one can't predict the future. This implies that something strange could happen whenever a star collapsed. We wouldn't be affected by the breakdown of prediction if the singularities are not naked – that is, they are not shielded from the outside. Penrose proposed the cosmic censorship conjecture: all singularities formed by the collapse of stars or other bodies are hidden from view inside black holes. A black hole is a region where gravity is so strong that light cannot escape. The cosmic censorship conjecture is almost certainly true, because a number of attempts to disprove it have failed.

When John Wheeler introduced the term 'black hole' in 1967, it replaced the earlier name of 'frozen star'. Wheeler's coinage emphasised that the remnants of collapsed stars are of interest in their own right, independently of how they were formed. The new name caught on quickly.

From the outside, you can't tell what is inside a black hole. Whatever you throw in, or however it is formed, black holes look the same. John Wheeler is known for expressing this principle as 'A black hole has no hair.'

A black hole has a boundary called the event horizon. It is where gravity is just strong enough to drag light back and prevent it from escaping. Because nothing can travel faster than light, everything else will get dragged back also. Falling through the event horizon is a bit like going over Niagara Falls in a canoe. If you are above the Falls, you can get away if you paddle fast enough, but once you are over the edge you are lost. There's no way back. As you get nearer the Falls, the current gets faster. This means it pulls harder on the front of the canoe than the back. There's a danger that the canoe will be pulled apart. It is the same with black holes. If you fall towards a black hole feet first, gravity will pull harder on your feet than your head, because they are nearer the black hole. The result is that you will be stretched out lengthwise, and squashed in sideways. If the black hole has a mass of a few times our Sun, you would be torn apart and made into spaghetti before you reached the horizon. However, if you fell into a much larger black hole, with a mass of more than a million times the Sun, the gravitational pull would be the same on the whole of your body and you would reach the horizon without difficulty. So, if you want to explore the inside of a black hole, make sure you choose a big one. There is a black hole with a mass of about four million times that of the Sun at the centre of our Milky Way galaxy.

Although you wouldn't notice anything in particular as you fell into a black hole, someone watching you from a distance would never see you cross the event horizon. Instead, you would appear to slow down and hover just outside. Your image would get dimmer and dimmer, and redder and redder, until you were effectively lost from sight. As far as the outside world is concerned, you would be lost for ever.

Shortly after the birth of my daughter Lucy I had a eureka moment. I discovered the area theorem. If general relativity is correct, and the energy density of matter is positive as is usually the case, then the surface area of the event horizon, the boundary of a black hole, has the property that it always increases when additional matter or radiation falls into the black hole. Moreover, if two black holes collide and merge to form a single black hole, the area of the event horizon around the resulting black hole is greater than the sum of the areas of the event horizons around the original black holes. The area theorem can be tested experimentally by the Laser Interferometer Gravitational-Wave Observatory (LIGO). On 14 September 2015, LIGO detected gravitational waves from the collision and merger of two black holes. From the waveform, one can estimate the masses and angular momenta of the black holes, and by the no-hair theorem these determine the horizon areas.

These properties suggest that there is a resemblance between the area of the event horizon of a black hole and conventional classical physics, specifically the concept of entropy in thermodynamics. Entropy can be regarded as a measure of the disorder of a system, or equivalently as a lack of knowledge of its precise state. The famous second law of thermodynamics says that entropy always increases with time. This discovery was the first hint of this crucial connection.

The analogy between the properties of black holes and the laws of thermodynamics can be extended. The first law of thermodynamics says that a small change in the entropy of a system is accompanied by a proportional change in the energy of the system. Brandon Carter, Jim Bardeen and I found a similar law relating the change in mass of a black hole to a change in the area of the event horizon. Here the factor of proportionality involves a quantity called the surface gravity, which is a measure of the strength of the gravitational field at the event horizon. If one accepts that the area of the event horizon is analogous to entropy, then it would seem that the surface gravity is analogous to temperature. The resemblance is strengthened by the fact that the surface gravity turns out to be the same at all points on the event horizon, just as the temperature is the same everywhere in a body at thermal equilibrium.

Although there is clearly a similarity between entropy and the area of the event horizon, it was not obvious to us how the area could be identified as the entropy of a black hole itself. What would be meant by the entropy of a black hole? The crucial suggestion was made in 1972 by Jacob Bekenstein, who was a graduate student at Princeton University. It goes like this. When a black hole is created by gravitational collapse, it rapidly settles down to a stationary state, which is characterised by three parameters: the mass, the angular momentum and the electric charge.

This makes it look as if the final black hole state is independent of whether the body that collapsed was composed of matter or antimatter, or whether it was spherical or highly irregular in shape. In other words, a black hole of a given mass, angular momentum and electric charge could have been formed by the collapse of any one of a large number of different configurations of matter. So what appears to be the same black hole could be formed by the collapse of a large number of different types of star. Indeed, if quantum effects are neglected, the number of configurations would be infinite since the black hole could have been formed by the collapse of a cloud of an indefinitely large number of particles of indefinitely low mass. But could the number of configurations really be infinite?

Quantum mechanics famously involves the uncertainty principle. This states that it is impossible to measure both the position and speed of any object. If one measures exactly where something is, then its speed is undetermined. If one measures the speed of something, then its position is undetermined. In practice, this means that it is impossible to localise anything. Suppose you want to measure the size of something, then you need to figure out where the ends of this moving object are. You can never do this accurately, because it will involve making a measurement of both the positions of something and its speed at the same time. In turn, it is then impossible to determine the size of an object. All you can do is to say that the uncertainty principle makes it impossible to say precisely what the size of something really is. It turns out that the uncertainty principle imposes a limit on the size of something. After a little bit of calculation, one finds that for a given mass of an object, there is a minimum size. This minimum size is small for heavy objects, but as one looks at lighter and lighter objects, the minimum size gets bigger and bigger. This minimum size can be thought of as a consequence of the fact that in quantum mechanics objects can be thought of either as a wave or a particle. The lighter an object is, the longer its wavelength is and so it is more spread out. The heavier an object is, the shorter its wavelength and so it will seem more compact. When these ideas are combined with those of general relativity, it means that only objects heavier than a particular weight can form black holes. That weight is about the same as that of a grain of salt. A further consequence of these ideas is that the number of configurations that could form a black hole of a given mass, angular momentum, and electric charge, although very large, may also be finite. Jacob Bekenstein suggested that from this finite number, one could interpret the entropy of a black hole. This would be a measure of the amount of information that seems irretrievably lost, during the collapse when a black hole was created.

The apparently fatal flaw in Bekenstein's suggestion was that, if a black hole has a finite entropy that is proportional to the area of its event horizon, it also ought to have a non-zero temperature which would be proportional to its surface gravity. This would imply that a black hole could be in equilibrium with thermal radiation at some temperature other than zero. Yet according to classical concepts no such equilibrium is possible since the black hole would absorb any thermal radiation that fell on it but by definition would not be able to emit anything in return. It cannot emit anything, it cannot emit heat.

This created a paradox about the nature of black holes, the incredibly dense objects created by the collapse of stars. One theory suggested that black holes with identical qualities could be formed from an infinite number of different types of stars. Another suggested that the number could be finite. This is a problem of information – the idea that every particle and every force in the universe contains information.

Because black holes have no hair, as the scientist John Wheeler put it, one can't tell from the outside what is inside a black hole, apart from its mass, electric charge and rotation. This means that a black hole must contain a lot of information that is hidden from the outside world. But there is a limit to the amount of information one can pack into a region of space. Information requires energy, and energy has mass by Einstein's famous equation, $E = mc^2$. So, if there's too much information in a region of space, it will collapse into a black hole, and the size of the black hole will reflect the amount of information. It is like piling more and more books into a library. Eventually, the shelves will give way and the library will collapse into a black hole.

If the amount of hidden information inside a black hole depends on the size of the hole, one would expect from general principles that the black hole would have a temperature and would glow like a piece of hot metal.

But that was impossible because, as everyone knew, nothing could get out of a black hole. Or so it was thought.

This problem remained until early in 1974, when I was investigating what the behaviour of matter in the vicinity of a black hole would be according to quantum mechanics. To my great surprise, I found that the black hole seemed to emit particles at a steady rate. Like everyone else at that time, I accepted the dictum that a black hole could not emit anything. I therefore put quite a lot of effort into trying to get rid of this embarrassing effect. But the more I thought about it, the more it refused to go away, so that in the end I had to accept it. What finally convinced me it was a real physical process was that the outgoing particles have a spectrum that is precisely thermal. My calculations predicted that a black hole creates and emits particles and radiation, just as if it were an ordinary hot body, with a temperature that is proportional to the surface gravity and inversely proportional to the mass. This made the problematic suggestion of Jacob Bekenstein, that a black hole had a finite entropy, fully consistent, since it implied that a black hole could be in thermal equilibrium at some finite temperature other than zero.

Since that time, the mathematical evidence that black holes emit thermal radiation has been confirmed by a number of other people with various different approaches. One way to understand the emission is as follows. Quantum mechanics implies that the whole of space is filled with pairs of virtual particles and antiparticles that are constantly materialising in pairs, separating and then coming together again, and annihilating each other. These particles are called virtual, because, unlike real particles, they cannot be observed directly with a particle detector. Their indirect effects can nonetheless be measured, and their existence has been confirmed by a small shift, called the Lamb shift, which they produce in the spectrum energy of light from excited hydrogen atoms. Now, in the presence of a black hole, one member of a pair of virtual particles may fall into the hole, leaving the other member without a partner with which to engage in mutual annihilation. The forsaken particle or antiparticle may fall into the black hole after its partner, but it may also escape to infinity, where it appears to be radiation emitted by the black hole.

Another way of looking at the process is to regard the member of the pair of particles that falls into the black hole, the antiparticle say, as being really a particle that is travelling backwards in time. Thus the antiparticle falling into the black hole can be regarded as a particle coming out of the black hole but travelling backwards in time. When the particle reaches the point at which the particle–antiparticle pair originally materialised, it is scattered by the gravitational field, so that it travels forward in time. A black hole of the mass of the Sun would leak particles at such a slow rate that it would be impossible to detect. However, there could be much smaller mini black holes with the mass of, say, a mountain. These might have formed in the very early universe if it had been chaotic and irregular. A mountain-sized black hole would give off X-rays and gamma rays, at a rate of about ten million megawatts, enough to power the world's electricity supply. It wouldn't be easy, however, to harness a mini black hole. You couldn't keep it in a power station because it would drop through the floor and end up at the centre of the Earth. If we had such a black hole, about the only way to keep hold of it would be to have it in orbit around the Earth.

People have searched for mini black holes of this mass, but have so far not found any. This is a pity because, if they had, I would have got a Nobel Prize. Another possibility, however, is that we might be able to create micro black holes in the extra dimensions of space-time. According to some theories, the universe we experience is just a four-dimensional surface in a ten- or eleven-dimensional space. The movie *Interstellar* gives some idea of what this is like. We wouldn't see these extra dimensions, because light wouldn't propagate through them but only through the four dimensions of our universe. Gravity, however, would affect the extra dimensions, and would be much stronger than in our universe. This would make it much easier to form a little black hole in the extra dimensions. It might be possible to observe this at the LHC, the Large Hadron Collider, at CERN in Switzerland. This consists of a circular tunnel, twenty-seven kilometres long. Two beams of particles travel round this tunnel in opposite directions and are made to collide. Some of the collisions might create micro black holes. These would radiate particles in a pattern that would be easy to recognise. So I might get a Nobel Prize after all.*

As particles escape from a black hole, the hole will lose mass and shrink. This will increase the rate of emission of particles. Eventually, the black hole will lose all its mass and disappear. What then happens to all the particles and unlucky astronauts that fell into the black hole? They can't just

---

* Nobel Prizes cannot be awarded posthumously and so sadly this ambition will now never be realised.

re-emerge when the black hole disappears. The particles that come out of a black hole seem to be completely random and to bear no relation to what fell in. It appears that the information about what fell in is lost, apart from the total amount of mass and the amount of rotation. But if information is lost, this raises a serious problem that strikes at the heart of our understanding of science. For more than 200 years, we have believed in scientific determinism – that is, that the laws of science determine the evolution of the universe.

If information were really lost in black holes, we wouldn't be able to predict the future, because a black hole could emit any collection of particles. It could emit a working television set or a leather-bound volume of the complete works of Shakespeare, though the chance of such exotic emissions is very low. It is much more likely to emit thermal radiation, like the glow from red-hot metal. It might seem that it wouldn't matter very much if we couldn't predict what comes out of black holes. There aren't any black holes near us. But it is a matter of principle. If determinism, the predictability of the universe, breaks down with black holes, it could break down in other situations. There could be virtual black holes that appear as fluctuations out of the vacuum, absorb one set of particles, emit another and disappear into the vacuum again. Even worse, if determinism breaks down, we can't be sure of our past history either. The history books and our memories could just be illusions. It is the past that tells us who we are. Without it, we lose our identity.

It was therefore very important to determine whether information really was lost in black holes, or whether in principle it could be recovered. Many scientists felt that information should not be lost, but for years no one suggested a mechanism by which it could be preserved. This apparent loss of information, known as the information paradox, has troubled scientists for the last forty years, and still remains one of the biggest unsolved problems in theoretical physics.

Recently, interest in possible resolutions of the information paradox has been revived as new discoveries have been made about the unification of gravity and quantum mechanics. Central to these recent breakthroughs is the understanding of the symmetries of space-time.

Suppose there was no gravity and space-time was completely flat. This would be like a completely featureless desert. Such a place has two types of symmetry. The first is called translation symmetry. If you moved from one point in the desert to another, you would not notice any change. The second symmetry is rotation symmetry. If you stood somewhere in the desert and started to turn around, you would again not notice any difference in what you saw. These symmetries are also found in 'flat' space-time, the space-time one finds in the absence of any matter.

If one put something into this desert, these symmetries would be broken. Suppose there was a mountain, an oasis and some cacti in the desert, it would look different in different places and in different directions. The same is true of space-time. If one puts objects into a space-time, the translational and rotational symmetries get broken. And introducing objects into a space-time is what produces gravity.

A black hole is a region of space-time where gravity is strong, space-time is violently distorted and so one expects its symmetries to be broken. However, as one moves away from the black hole, the curvature of space-time gets less and less. Very far away from the black hole, space-time looks very much like flat space-time.

Back in the 1960s, Hermann Bondi, A. W. Kenneth Metzner, M. G. J. van der Burg and Rainer Sachs made the truly remarkable discovery that space-time far away from any matter has an infinite collection of symmetries known as supertranslations. Each of these symmetries is associated with a conserved quantity, known as the supertranslation charges, A conserved quantity is a quantity that does not change as a system evolves. These are generalisations of more familiar conserved quantities. For example, if space-time does not change in time, then energy is conserved. If space-time looks the same at different points in space, then momentum is conserved.

What was remarkable about the discovery

of supertranslations is that there are an infinite number of conserved quantities far from a black hole. It is these conservation laws that have given an extraordinary and unexpected insight into process in gravitational physics.

In 2016, together with my collaborators Malcolm Perry and Andy Strominger, I have been working on using these new results with their associated conserved quantities to find a possible resolution to the information paradox. We know that the three discernible properties of black holes are their mass, their charge and their angular momentum. These are the classical charges that have been understood for a long time. However, black holes also carry a supertranslation charge. So perhaps black holes have a lot more to them than we first thought. They are not bald or with only three hairs, but actually have a very large amount of supertranslation hair.

This supertranslation hair might encode some of the information about what is inside the black hole. It is likely that these supertranslation charges do not contain all of the information, but the rest might be accounted for by some additional conserved quantities, superrotation charges, associated with some additional related symmetries called superrotations, which are, as yet, not well understood. If this is right, and all the information about a black hole can be understood in terms of its 'hairs', then perhaps there is no loss of information. These ideas have just received confirmation with our most recent calculations. Strominger, Perry and myself, together with a graduate student, Sasha Haco, have discovered that these superrotation charges an account for the entire entropy of any black hole. Quantum mechanics continues to hold, and information is stored on the horizon, the surface of the black hole.

The black holes are still characterised only by their overall mass, electric charge and spin outside the event horizon but the event horizon itself contains the information needed to tell us about what has fallen into the black hole in a way that goes beyond these three characteristics the black hole has. People are still working on these issues and therefore the information paradox remains unresolved. But I am optimistic that we are moving towards a solution. Watch this space.

IF YOU FEEL
YOU ARE IN
A BLACK HOLE,
DON'T GIVE UP.
THERE'S A WAY OUT

IS TIME TRAVEL POSSIBLE?

IN science fiction, space and time warps are commonplace. They are used for rapid journeys around the galaxy or for travel through time. But today's science fiction is often tomorrow's science fact. So what are the chances of time travel?

The idea that space and time can be curved or warped is fairly recent. For more than 2,000 years the axioms of Euclidean geometry were considered to be self-evident. As those of you who were forced to learn geometry at school may remember, one of the consequences of these axioms is that the angles of a triangle add up to 180 degrees.

However, in the last century people began

to realise that other forms of geometry were possible in which the angles of a triangle need not add up to 180 degrees. Consider for example the surface of the Earth. The nearest thing to a straight line on the surface of the Earth is what is called a great circle. These are the shortest paths between two points so they are the routes that airlines use. Consider now the triangle on the surface of the Earth made up of the equator, the line of 0 degrees longitude through London and the line of 90 degrees longtitude east through Bangladesh. The two lines of longitude meet the equator at a right angle or 90 degrees. The two lines of longitude also meet each other at the North Pole at a right angle or 90 degrees. Thus one has a triangle with three right angles. The angles of this triangle add up to 270 degrees which is obviously greater than the 180 degrees for a triangle on a flat surface. If one drew a triangle on a saddle-shaped surface one would find that the angles added up to less than 180 degrees.

The surface of the Earth is what is called a two-dimensional space. That is, you can move on the surface of the Earth in two directions at right angles to each other: you can move north–south or east–west. But of course there is a third direction at right angles to these two and that is up or down. In other words the surface of the Earth exists in three-dimensional space. The three-dimensional space is flat. That is to say it obeys Euclidean geometry. The angles of a triangle add up to 180 degrees. However, one could imagine a race of two-dimensional creatures who could move about on the surface of the Earth but who couldn't experience the third direction of up or down. They wouldn't know about the flat three-dimensional space in which the surface of the Earth lives. For them space would be curved and geometry would be non-Euclidean.

But just as one can think of two-dimensional beings living on the surface of the Earth so one could imagine that the three-dimensional space in which we live was the surface of a sphere in another dimension that we don't see. If the sphere were very large, space would be nearly flat and Euclidean geometry would be a very good approximation over small distances. But we would notice that Euclidean geometry broke down over large distances. As an illustration of this imagine a team of painters adding paint to the surface of a large ball.

As the thickness of the paint layer increased, the surface area would go up. If the ball were in a flat three-dimensional space one could go on adding paint indefinitely and the ball would get bigger and bigger. However, if the three-dimensional space were really the surface of a sphere in another dimension its volume would be large but finite. As one added more layers of paint the ball would eventually fill half the space. After that the painters would find that they were trapped in a region of ever-decreasing size, and almost the whole of space would be occupied by the ball and its layers of paint. So they would know that they were living in a curved space and not a flat one.

This example shows that one cannot deduce the geometry of the world from first principles as the ancient Greeks thought. Instead one has to measure the space we live in and find out its geometry by experiment. However, although a way to describe curved spaces was developed by the German Bernhard Riemann in 1854, it remained just a piece of mathematics for sixty years. It could describe curved spaces that existed in the abstract but there seemed no reason why the physical space we lived in should be curved. This reason came only in 1915 when Einstein put forward the general theory of relativity.

General relativity was a major intellectual revolution that has transformed the way we think about the universe. It is a theory not only of curved space but of curved or warped time as well. Einstein had realised in 1905 that space and time are intimately connected with each other, which is when his theory of special relativity was born, relating space and time to each other. One can describe the location of an event by four numbers. Three numbers describe the position of the event. They could be miles north and east of Oxford Circus and the height above sea level. On a larger scale they could be galactic latitude and longitude and distance from the centre of the galaxy.

The fourth number is the time of the event.

Thus one can think of space and time together as a four-dimensional entity called space-time. Each point of space-time is labelled by four numbers that specify its position in space and in time. Combining space and time into space-time in this way would be rather trivial if one could disentangle them in a unique way. That is to say if there was a unique way of defining the time and position of each event. However, in a remarkable paper written in 1905 when he was a clerk in the Swiss patent office, Einstein showed that the time and position at which one thought an event occurred depended on how one was moving. This meant that time and space were inextricably bound up with each other.

The times that different observers would assign to events would agree if the observers were not moving relative to each other. But they would disagree more the faster their relative speed. So one can ask how fast does one need to go in order that the time for one observer should go backwards relative to the time of another observer. The answer is given in the following limerick:

There was a young lady of Wight
Who travelled much faster than light
She departed one day
In a relative way
And arrived on the previous night.

So all we need for time travel is a spaceship that will go faster than light. Unfortunately in the same paper Einstein showed that the rocket power needed to accelerate a spaceship got greater and greater the nearer it got to the speed of light. So it would take an infinite amount of power to accelerate past the speed of light.

Einstein's paper of 1905 seemed to rule

out time travel into the past. It also indicated that space travel to other stars was going to be a very slow and tedious business. If one couldn't go faster than light the round trip to the nearest star would take at least eight years and to the centre of the galaxy at least 80,000 years. If the spaceship went very near the speed of light it might seem to the people on board that the trip to the galactic centre had taken only a few years. But that wouldn't be much consolation if everyone you had known had died and been forgotten thousands of years ago when you got back. That wouldn't be much good for science-fiction novels either, so writers had to look for ways to get round this difficulty.

In 1915, Einstein showed that the effects of gravity could be described by supposing that space-time was warped or distorted by the matter and energy in it, and this theory is known as general relativity. We can actually observe this warping of space-time produced by the mass of the Sun in the slight bending of light or radio waves passing close to the Sun.

This causes the apparent position of the star or radio source to shift slightly when the Sun is between the Earth and the source. The shift is very small, about a thousandth of a degree, equivalent to a movement of an inch at a distance of a mile. Nevertheless it can be measured with great accuracy and it agrees with the predictions of general relativity. We have experimental evidence that space and time are warped.

The amount of warping in our neighbourhood is very small because all the gravitational fields in the solar system are weak. However we know that very strong fields can occur for example in the Big Bang or in black holes. So can space and time be warped enough to meet the demands from science fiction for things like hyperspace drives, wormholes or time travel? At first sight all these seem possible. For example, in 1948 Kurt Gödel found a solution to Einstein's field equations of general relativity that represents a universe in which all the matter was rotating. In this universe it would be possible to go off in a spaceship and come back before you had set out. Gödel was at the Institute of Advanced Study in Princeton where Einstein also spent his last years. He was more famous for proving you couldn't prove everything that is true even in such an apparently simple subject as arithmetic. But what he proved about general relativity allowing time travel really upset Einstein, who had thought it wouldn't be possible.

We now know that Gödel's solution couldn't represent the universe in which we live because it was not expanding. It also had a fairly large value for a quantity called the cosmological constant which is generally believed to be very small. However, other apparently more reasonable solutions that allow time travel have since been found. A particularly interesting one from an approach known as string theory contains two cosmic strings moving past each other at a speed very near to but slightly less than the speed of light. Cosmic strings are a remarkable idea of theoretical physics which science-fiction writers don't really seem to have caught on to. As their name suggests they are like string in that they have length but a tiny cross-section. Actually they are more like rubber bands because they are under enormous tension, something like a hundred billion billion billion tonnes. A cosmic string attached to the Sun would accelerate it from nought to sixty in a thirtieth of a second.

Cosmic strings may sound far fetched and pure science fiction, but there are good scientific reasons to believe they could have formed in the very early universe shortly after the Big Bang. Because they are under such great tension one might have expected them to accelerate to almost the speed of light.

What both the Gödel universe and the fast-moving cosmic-string space-time have in common is that they start out so distorted and curved that space-time curves back on itself and travel into the past was always possible. God might have created such a warped universe, but we have no reason to think that he did. All the evidence is that the universe started out in the Big Bang without the kind of warping needed to allow travel into the past. Since we can't change the way the universe began, the question of whether time travel is possible is one of whether we can subsequently make space-time so warped that one can go back to the past. I think this is an important subject for research, but one has to be careful not to be labelled a crank. If one made a research grant application to work on time travel it would be dismissed immediately. No government agency could afford to be seen to be spending public money on anything as way out as time travel. Instead one has to use technical terms like closed time-like curves which are code for time travel. Yet it is a very serious question. Since general relativity can permit time travel, does it allow it in our universe? And if not, why not?

Closely related to time travel is the ability to travel rapidly from one position in space to another. As I said earlier, Einstein showed that it would take an infinite amount of rocket power to accelerate a spaceship to beyond the speed of light. So the only way to get from one side of the galaxy to the other in a reasonable time would seem to be if we could warp space-time so much that we created a little tube or wormhole. This could connect the two sides of the galaxy and act as a short cut to get from one to the other and back while your friends were still alive. Such wormholes have been seriously suggested as being within the capabilities of a future civilisation. But if you can travel from one side of the galaxy to the other in a week or two you could go back through another wormhole and arrive back before you had set out. You could even manage to travel back in time with a single wormhole if its two ends were moving relative to each other.

One can show that to create a wormhole one needs to warp space-time in the opposite way to that in which normal matter warps it. Ordinary matter curves space-time back on itself like the surface of the Earth. However, to create a wormhole one needs matter that warps space-time in the opposite way like the surface of a saddle. The same is true of any other way of warping space-time to allow travel to the past if the universe didn't begin so warped that it allowed time travel. What one would need would be matter with negative mass and negative energy density to make space-time warp in the way required.

Energy is rather like money. If you have a positive bank balance, you can distribute it in various ways. But, according to the classical laws that were believed until quite recently, you weren't allowed to have an energy overdraft. So these classical laws would have ruled out us being able to warp the universe in the way required to allow time travel. However, the classical laws were overthrown by quantum theory, which is the other great revolution in our picture of the universe apart from general relativity. Quantum theory is more relaxed and allows you to have an overdraft on one or two accounts. If only the banks were as accommodating. In other words, quantum theory allows the energy density to be negative in some places provided it is positive in others.

The reason quantum theory can allow the energy density to be negative is that it is based on the Uncertainty Principle. This says that certain quantities like the position and speed of a particle can't both have well-defined values. The more accurately the position of a particle is defined the greater is the uncertainty in its speed, and vice versa. The Uncertainty Principle also applies to fields like the electromagnetic field or the gravitational field. It implies that these fields can't be exactly zero even in what we think of as empty space. For if they were exactly zero their values would have both a well-defined position at zero and a well-defined speed which was also zero. This would be a violation of the Uncertainty Principle. Instead the fields would have to have a certain minimum amount of fluctuations. One can interpret these so-called vacuum fluctuations as pairs of particles and antiparticles that suddenly appear together, move apart and then come back together again and annihilate each other.

These particle–antiparticle pairs are said to be virtual because one cannot measure them directly with a particle detector. However, one can observe their effects indirectly. One way of doing this is by what is called the Casimir effect. Imagine that you have two parallel metal plates a short distance apart. The plates act like mirrors for the virtual particles and anti-particles. This means that the region between the plates is a bit like an organ pipe and will only admit light waves of certain resonant frequencies. The result is that there are a slightly different number of vacuum fluctuations or virtual particles between the plates than there are outside them where vacuum fluctuations can have any wavelength. The difference in the number of virtual particles between the plates compared with outside the plates means that they don't exert as much pressure on one side of the plates compared with the other. There is thus a slight force pushing the plates together. This force has been measured experimentally. So, virtual particles actually exist and produce real effects.

Because there are fewer virtual particles or vacuum fluctuations between the plates, they have a lower energy density than in the region outside. But the energy density of empty space far away from the plates must be zero. Otherwise it would warp space-time and the universe wouldn't be nearly flat. So the energy density in the region between the plates must be negative.

We thus have experimental evidence from the bending of light that space-time is curved and confirmation from the Casimir effect that we can warp it in the negative direction. So it might seem that as we advance in science and technology we might be able to construct a wormhole or warp space and time in some other way so as to be able to travel into our past. If this were the case it would raise a whole host of questions and problems. One of these is if time travel will be possible in the future, why hasn't someone come back from the future to tell us how to do it.

Even if there were sound reasons for keeping us in ignorance, human nature being what it is it is difficult to believe that someone wouldn't show off and tell us poor benighted peasants the secret of time travel. Of course, some people would claim that we have already been visited from the future. They would say that UFOs come from the future and that governments are engaged in a gigantic conspiracy to cover them up and keep for themselves the scientific knowledge that these visitors bring. All I can say is that if governments were hiding something they are doing a poor job of extracting useful information from the aliens. I'm pretty sceptical of conspiracy theories, as I believe that cock-up theory is more likely. The reports of sightings of UFOs cannot all be caused by extra-terrestrials because they are mutually contradictory. But, once you admit that some are mistakes or hallucinations, isn't it more probable that they all are than that we are being visited by people from the future or from the other side of the galaxy? If they really want to colonise the Earth or warn us of some danger they are being rather ineffective.

A possible way to reconcile time travel with the fact that we don't seem to have had any visitors from the future would be to say that such travel can occur only in the future. In this view one would say space-time in our past was fixed because we have observed it and seen that it is not warped enough to allow travel into the past. On the other hand the future is open. So we might be able to warp it enough to allow time travel. But because we can warp space-time only in the future we wouldn't be able to travel back to the present time or earlier.

This picture would explain why we haven't been overrun by tourists from the future. But it would still leave plenty of paradoxes. Suppose it were possible to go off in a rocket ship and come back before you had set off. What would stop you blowing up the rocket on its launch pad or otherwise preventing yourself from setting out in the first place? There are other versions of this paradox, like going back and killing your parents before you were born, but they are essentially equivalent. There seem to be two possible resolutions.

One is what I shall call the consistent-histories approach. It says that one has to find a consistent solution of the equations of physics even if space-time is so warped that it is possible to travel into the past. On this view you couldn't set out on the rocket ship to travel into the past unless you had already come back and failed to blow up the launch pad. It is a consistent picture, but it would imply that we were completely determined: we couldn't change our minds. So much for free will.

The other possibility is what I call the alternative-histories approach. It has been championed by the physicist David Deutsch and it seems to have been what the creator of *Back to the Future* had in mind. In this view, in one alternative history there would not have been any return from the future before the rocket set off and so no possibility of it being blown up. But when the traveller returns from the future he enters another alternative history. In this the human race makes a tremendous effort to build a space-ship but just before it is due to be launched a similar spaceship appears from the other side of the galaxy and destroys it.

David Deutsch claims support for the alternative-histories approach from the sum-over-histories concept introduced by the physicist Richard Feynman. The idea is that according to quantum theory the universe doesn't just have a unique single history. Instead the universe has every single possible history, each with its own probability. There must be a possible history in which there is a lasting peace in the Middle East, though maybe the probability is low.

In some histories space-time will be so

warped that objects like rockets will be able to travel into their pasts. But each history is complete and self-contained, describing not only the curved space-time but also the objects in it. So a rocket cannot transfer to another alternative history when it comes round again. It is still in the same history which has to be self-consistent. Thus despite what Deutsch claims I think the sum-over-histories idea supports the consistent-histories hypothesis rather than the alternative-histories idea.

It thus seems that we are stuck with the consistent-histories picture. However, this need not involve problems with determinism or free will if the probabilities are very small for histories in which space-time is so warped that time travel is possible over a macroscopic region. This is what I call the Chronology Protection Conjecture: the laws of physics conspire to prevent time travel on a macro-scopic scale.

It seems that what happens is that when space-time gets warped almost enough to allow travel into the past virtual particles can almost become real particles following closed trajectories. The density of the virtual particles and their energy become very large. This means that the probability of these histories is very low. Thus it seems there may be a Chronology Protection Agency at work making the world safe for historians. But this subject of space and time warps is still in its infancy. According to string theory, which is our best hope of uniting general relativity and quantum theory, space-time ought to have ten dimensions, not just the four that we experience. The idea is that six of these ten dimensions are curled up into a space so small that we don't notice them. On the other hand the remaining four directions are fairly flat and are what we call space-time. If this picture is correct it might be possible to arrange that the four flat directions get mixed up with the six highly curved or warped directions. What this would give rise to we don't yet know. But it opens exciting possibilities.

In conclusion, rapid space travel and travel back in time can't be ruled out according to our present understanding. They would cause great logical problems, so let's hope there's a Chronology Protection Law to prevent people going back and killing our parents. But science-fiction fans need not lose heart. There's hope in string theory.